BABY SENSES

A Sensory Neuroscience Primer for All Ages

Illustrated and Written by

Dr. Jaya Viswanathan

Neuroscientist, Artist, Engineer, Educator

Archway Publishing books may be ordered through booksellers or by contacting:

Archway Publishing
1663 Liberty Drive
Bloomington, IN 47403
www.archwaypublishing.com
844-669-3957

Cover Image: A purkinje neuron wraps around illustrations of the animals featured in this book.
Purkinje neurons are some of the largest neurons found in humans, within the cerebellum,
where they contribute to the production of smooth, continuous, and accurate movements.

ISBN: 978-1-6657-3715-9 (sc)
ISBN: 978-1-6657-3716-6 (hc)
ISBN: 978-1-6657-3714-2 (e)

Print information available on the last page.

Archway Publishing rev. date: 04/11/2023

This book is dedicated to my incredible Mama and Papa, for instilling in me both an undying curiosity about the world around me and deep appreciation of how we relate to it.

INTRODUCTION

Life began as single-celled organisms, approximately 3.8 billion years ago. Although the precise sequence of events are not known, key steps included the formation of organic chemicals and their enclosure within a membrane separated from the rest of the world [1]. This separation between "inside" and "outside" is crucial to all living beings. The enclosure and all it contains – the cell – is the basic unit of life in all living things. From industrious ants to mischievous whales, sensing, understanding, and responding appropriately to the world around us is an essential function of life. Over millions of years, the nervous system has evolved to perform this crucial function – *interpreting the world* - in various animal species based on unique challenges they faced in different environments.

Since the agricultural revolution about 10,000 years ago, the human nervous system – and brain – began evolving differently. The need to execute complex tasks such as seasonal planting of crops necessitated that our ancestors plan far into the future, a cognitive ability that today distinguishes us from other known species. It was perhaps this behavioral demand that also expanded our cognitive development, capacity to build mental maps, and allowed us to become *consciously* curious about the world. In 1879, Wilhelm Wundt set up the first experimental psychology lab to investigate sensory perception and consciousness [2] and since then several psychologists, physicists, physiologists, biologists, neuroscientists, mathematicians, and computational scientists have been working on trying to understand how information from different sensory modalities are perceived in various animals.

Today, as modern humans, we enjoy music and art simply for their beauty [3]. As a neuroscientist and artist, I find the intricacies of neurological architectures and forms to be exquisite. This book is aimed at highlighting, to an intellectually curious audience of all ages, both the amazing ways in which various animals have evolved to interpret their worlds, as well as the art and science of the neural systems supporting these senses. The systems highlighted here span the traditional 5 – historically considered as the 5 human senses (vision, audition, olfaction, somatosensation, and gustation), internal sensory systems that help animals maintain **homeostasis** or internal stability, to more exotic systems such as those that enable animals to sense electrical and magnetic fields. Each animal was chosen as a unique exemplar for a sense – for instance, while all animals possess a sense of taste, certain fish rely on this sense exclusively, while humans enjoy the smells, sight, texture, and taste of food. More detailed information about the neuroscientific terminology and processes – highlighted in the dialogue - can be found in the expanded glossary. Names of various species and their juveniles are italicized in the text dialogue and are listed under further information.

The neuroscience paintings in this book illustrate the world of **sensory neuroscience** normally only accessible through a microscope to neuroscientists. While scientifically accurate, these are my personal and artistic interpretations of images sourced from textbooks and original research articles. These snapshots also document sensory systems at various levels of magnification, so while in some you can identify brain cells and muscles, in others, you can go even further and spot **intracellular organelles** such as mitochondria, known as the powerhouses of the cell. I hope this science-art book brings you a deeper appreciation of the amazing animals we share our planet with as well as the beauty of their nervous systems.

THE TRADITIONAL FIVE

"Papa! Papa!" said the baby *Lemur*,
"We can jump, play, and sing at night,
We do not seem to need our sight,
How do we live in the dark?"

"That's our superpower, *pup*" replied papa Lemur,
"When we smell things around us,
They go straight to our brain with each breath,
We have <u>*olfactory bulbs*</u> which map our world,
So we can find food, friends, and our way home!"

"That's our special skill, *hatchling*", replied mama Squid,
"We find special flashlights when we are young,
Tiny _bioluminescent_ buddies who light our way,
They help us hide using light and shadow,
So we keep our shiny friends safe in our mantle!"

5

"Mama! Mama!" said the baby *Tawny Owl*,
"To find food you don't need any light,
You sing duets with friends all night,
How will I do these when I'm older?"

6

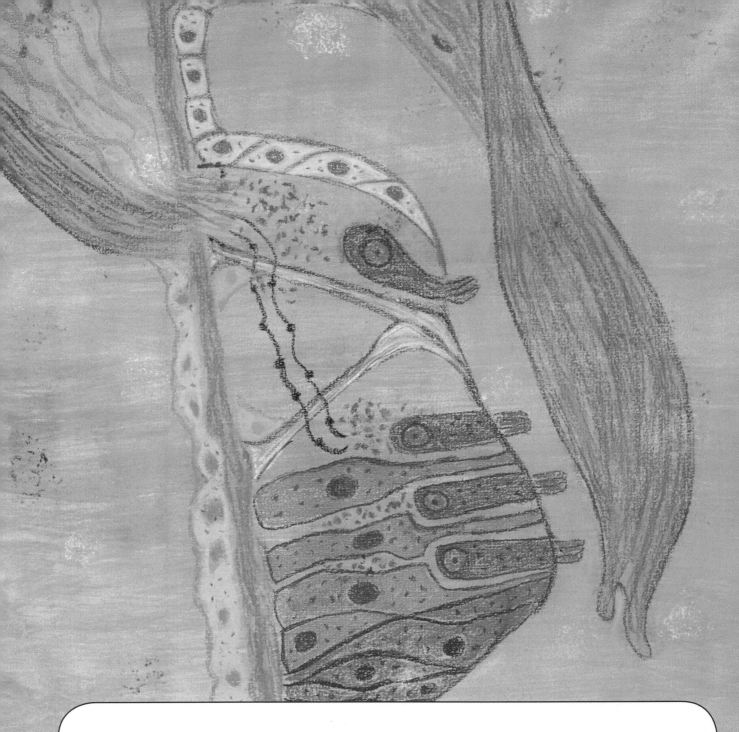

"That's our extraordinary ability, *owlet*", replied mama Owl,
"Our ears are hidden behind our big eyes,
So our super-hearing is often a surprise,
We precisely *localize sounds,* even faint rustles,
And hear our friends, near or far, when they whistle!"

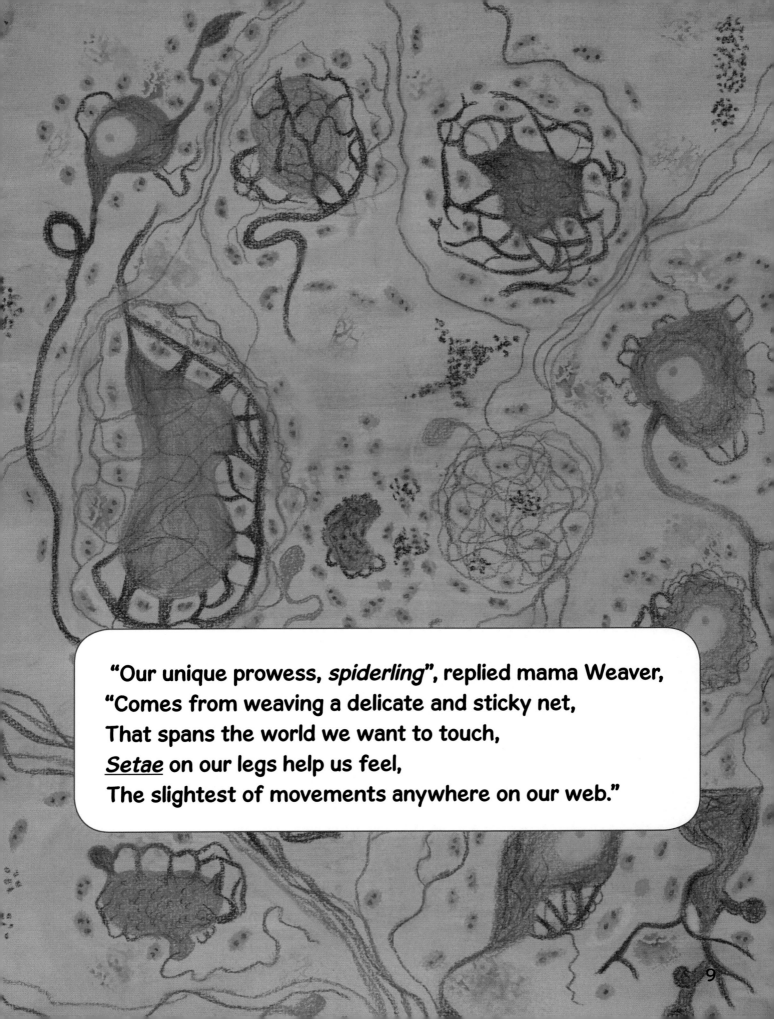

"Our unique prowess, *spiderling*", replied mama Weaver,
"Comes from weaving a delicate and sticky net,
That spans the world we want to touch,
<u>Setae</u> on our legs help us feel,
The slightest of movements anywhere on our web."

9

"Papa! Papa!" said the baby *Butterfly Goldfish*,
"I love to wave my fins and dance,
As I swim around our pond in a trance,
How do I know what not to eat?"

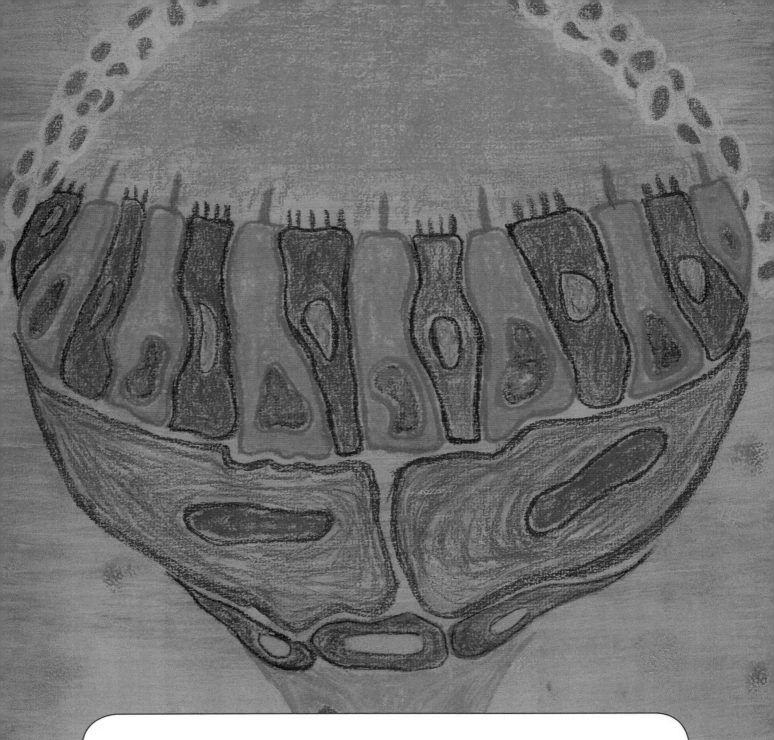

"Tasting is our superpower, *fry*" replied papa Goldfish,
"We taste our world with our whole body,
We have sensitive _taste buds_ everywhere,
They help us swim towards food,
And we get to try different flavors along the way!"

SENSING INNER WORLDS

"Papa! Papa!" said the baby *Emperor Penguin*,
"The tasty fish are in icy waters,
You dive so long while you can't breathe,
How do you find food for you and me?"

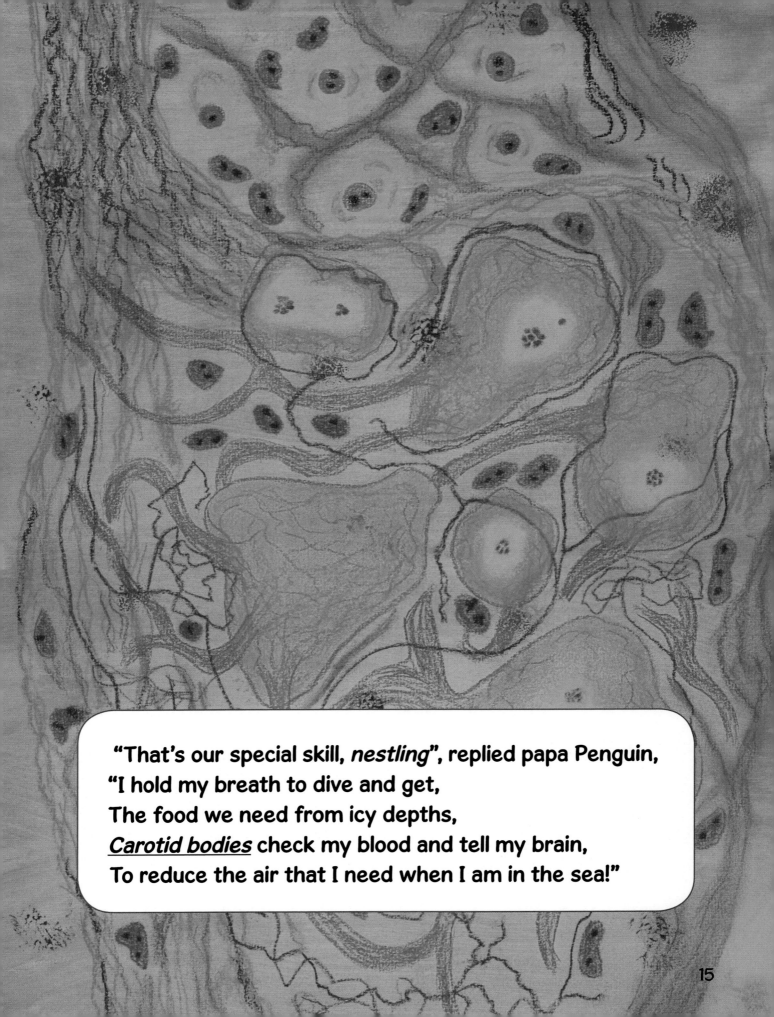

"That's our special skill, *nestling*", replied papa Penguin,
"I hold my breath to dive and get,
The food we need from icy depths,
<u>*Carotid bodies*</u> check my blood and tell my brain,
To reduce the air that I need when I am in the sea!"

15

"Mama! Mama!" said the baby *Langur*,
"Jumping from branch to branch is fun,
"I like to climb trees and run amok,
"How do I keep from falling?"

16

"That's our extraordinary ability, *infant*", replied mama Langur,
"*Neuromuscular spindles* keep checking our muscles,
And inform our brains how to move them,
So all our movements are exact,
As we climb, jump, run, and play!"

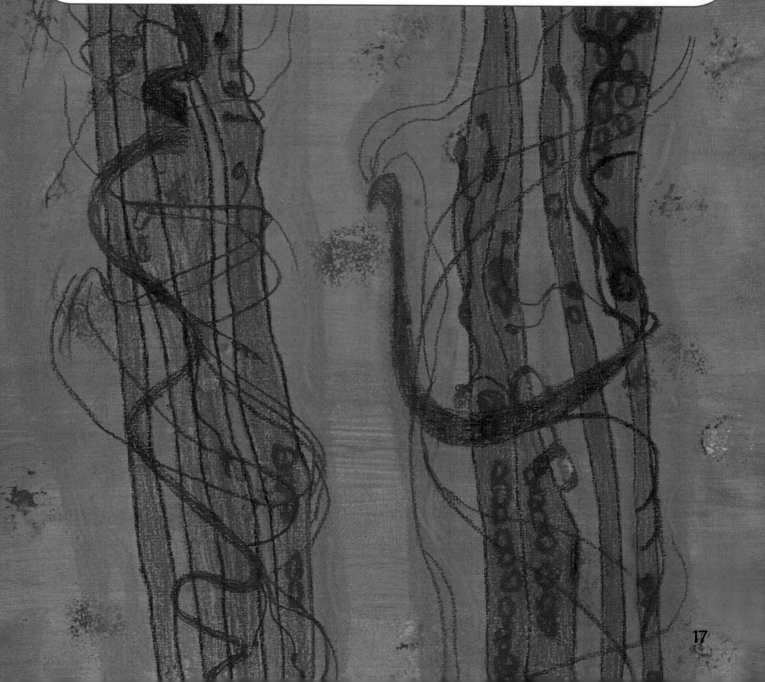

"Mama! Mama!" said the baby *Sloth*,
"The tops of trees are so much fun,
Hanging upside down all day in the sun,
But how do we not get dizzy?"

18

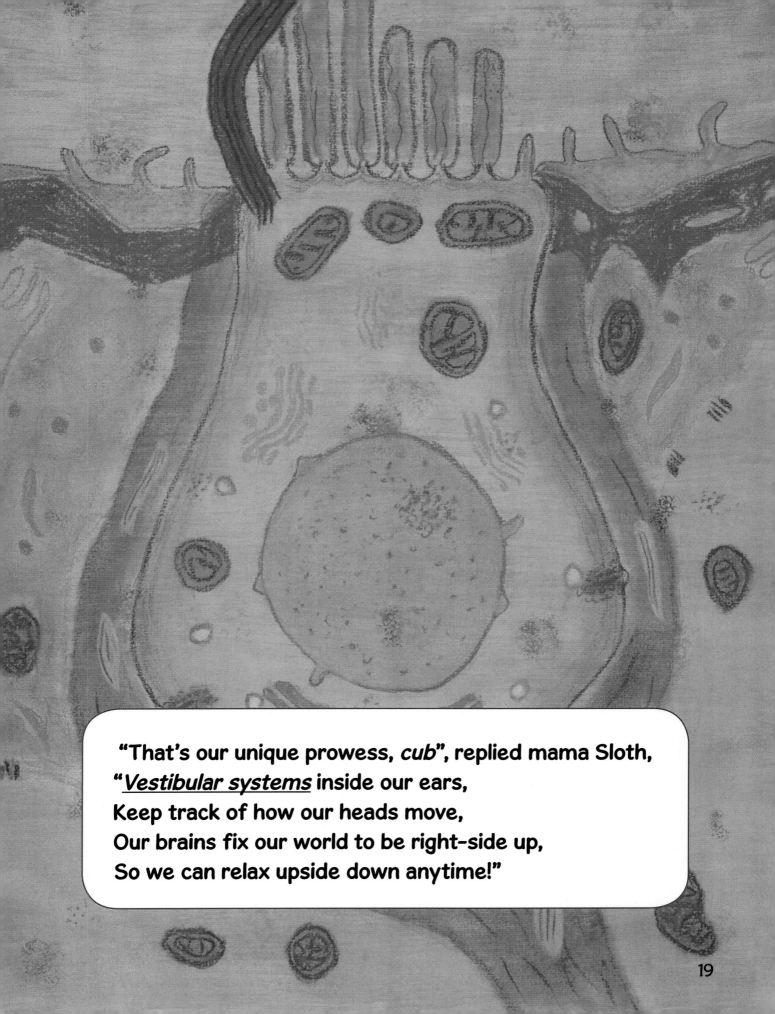

"That's our unique prowess, *cub*", replied mama Sloth,
"*Vestibular systems* inside our ears,
Keep track of how our heads move,
Our brains fix our world to be right-side up,
So we can relax upside down anytime!"

SPECIAL SENSES

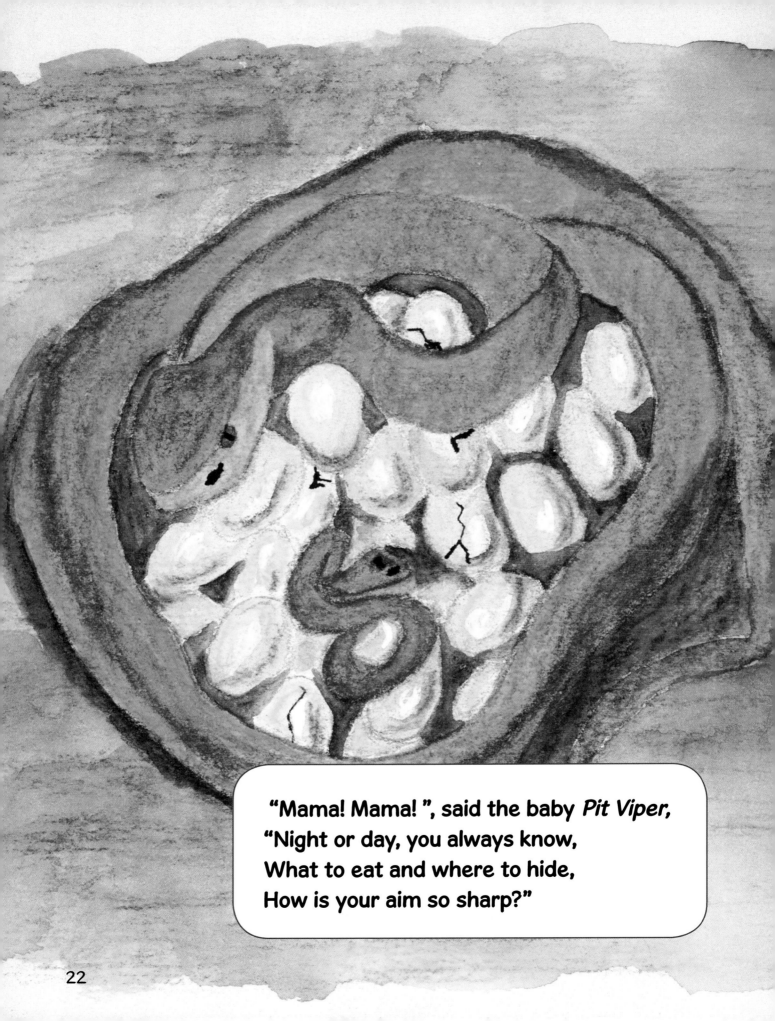

"Mama! Mama! ", said the baby *Pit Viper*,
"Night or day, you always know,
What to eat and where to hide,
How is your aim so sharp?"

"That's our superpower, *snakelet*", replied mama Viper,
"Our pit organs let us 'see' beyond light,
To *infrared radiation* which helps us find,
Warmth, food and shelter, especially at night,
We know where things are by 'seeing' their heat!"

"Mama! Mama!" said the baby *Worm*,
"I am so small compared to other animals,
But I can still hide, seek, and dig for treasure,
How can I do all these things?"

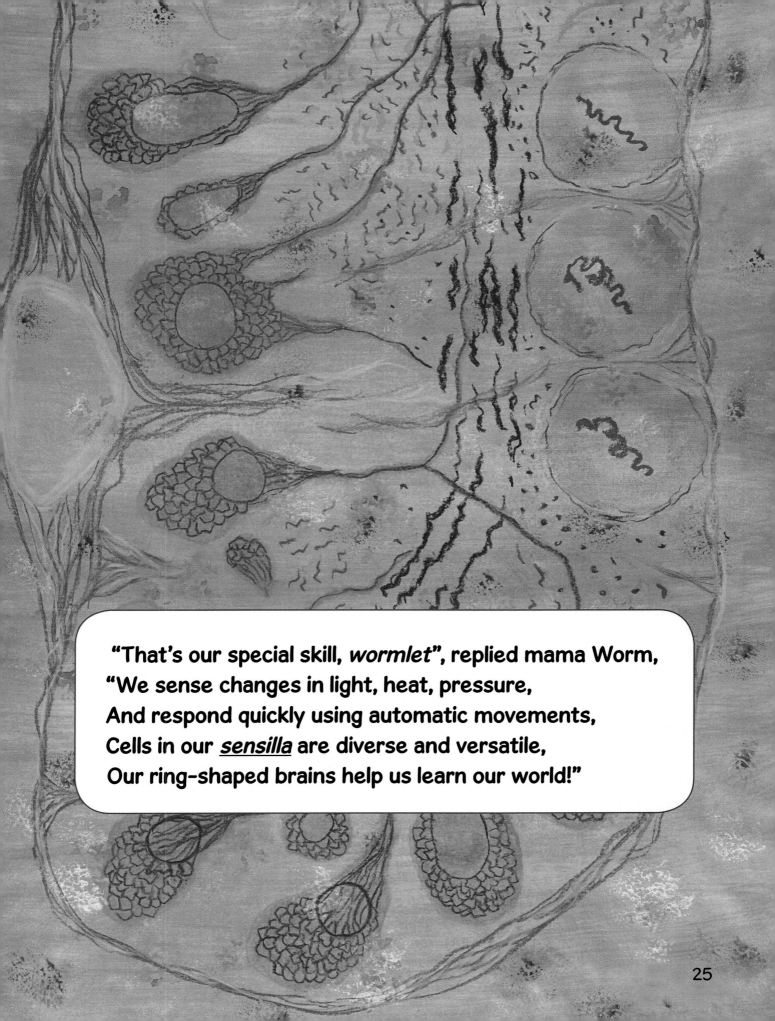

"That's our special skill, *wormlet*", replied mama Worm,
"We sense changes in light, heat, pressure,
And respond quickly using automatic movements,
Cells in our <u>sensilla</u> are diverse and versatile,
Our ring-shaped brains help us learn our world!"

"My Queen and Mama!" said the baby *Honeybee*,
"I am just one in a big, strong, community,
We all work together in perfect accord,
How will I know what my job will be?"

26

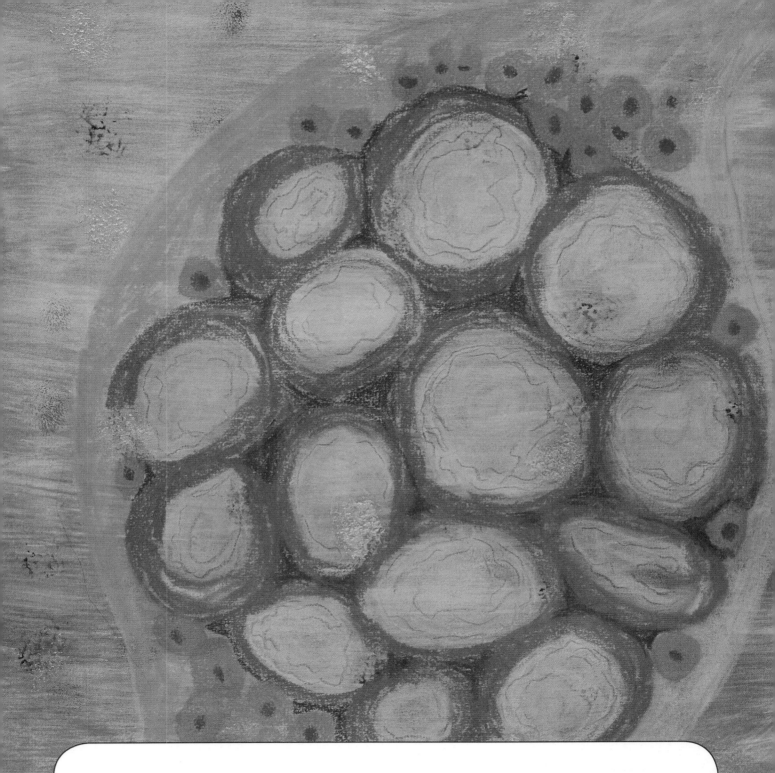

"That's our unique prowess, *larva*", replied queen Honeybee,
"We share work and fun in our large family,
Different or similar, we live in harmony,
<u>Pheromones</u> will tell you what tasks need to be done,
So that we can prosper together as a whole colony."

"Mama! Mama!" said the baby *Narwhal*,
"We live in cold arctic waters,
With ice and danger everywhere,
How do we stay safe?"

28

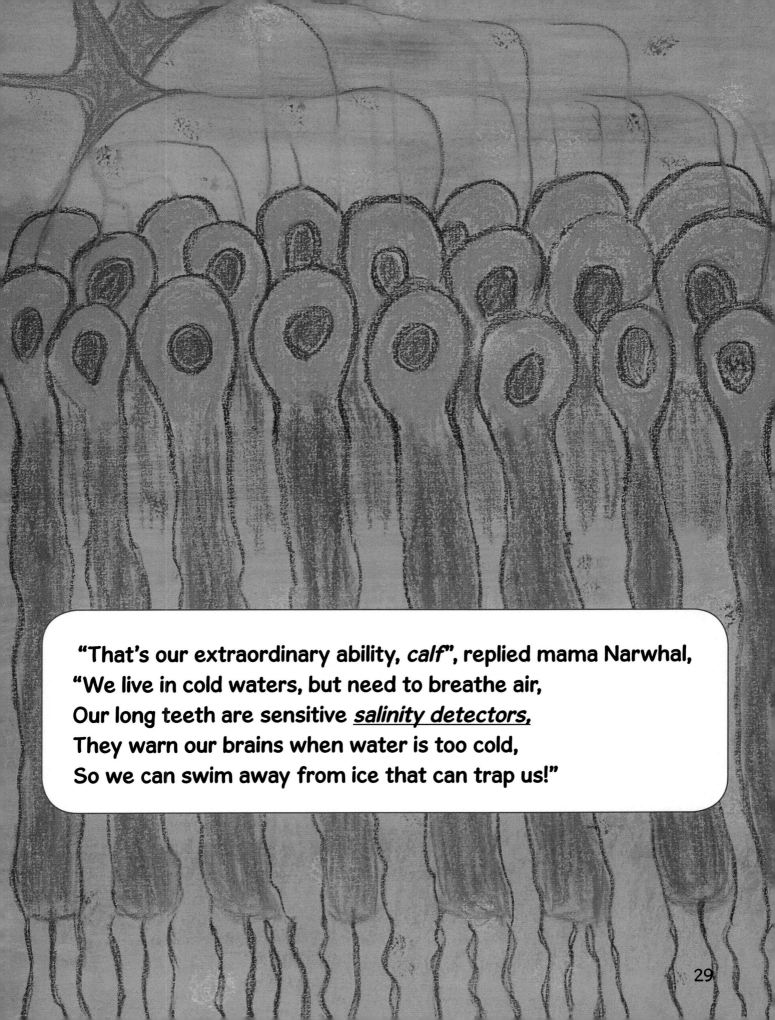

"That's our extraordinary ability, *calf*", replied mama Narwhal,
"We live in cold waters, but need to breathe air,
Our long teeth are sensitive _salinity detectors,_
They warn our brains when water is too cold,
So we can swim away from ice that can trap us!"

"Papa! Papa!" said the *Monarch* caterpillar,
"As a caterpillar all I do is eat and grow,
But you fly to far-away places to find warmth,
How do you find your way in the sky?"

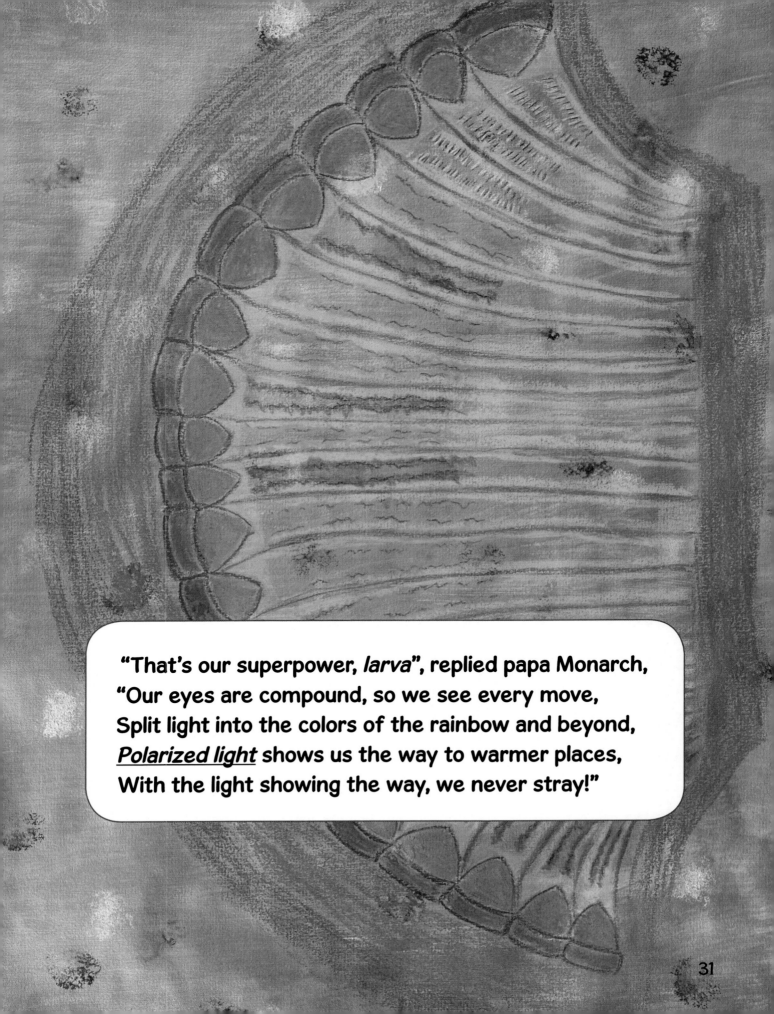

"That's our superpower, *larva*", replied papa Monarch,
"Our eyes are compound, so we see every move,
Split light into the colors of the rainbow and beyond,
Polarized light shows us the way to warmer places,
With the light showing the way, we never stray!"

"Mama! Mama!" said the baby *Blue spotted Ray*,
"We find our food when the water is murky,
Travel far and wide, never losing our way,
How do we explore the whole world?"

"That's our special skill, *pup*", replied mama Ray,
"All living things have some electricity,
And the poles of the world make it magnetic,
Our *ampullae of Lorenzini* help us know,
Electric and magnetic energy, and how they flow!"

"Papa! Papa!" said the baby *Poison Dart Frog,*
"Our skin lets us breathe on land or in water,
Our bright colors warn of the poison that protects us,
What else makes our skin special?"

"Our skin has unique prowess, *tadpole*", replied papa Frog,
"<u>Neuromasts</u> on it help us swim away from danger,
They tell us how the water flows around us,
They help us hear sounds when under water,
These abilities let our family survive millions of years!"

35

"Mama! Mama!" said the baby *Bat*,
"We stay where it is warm, damp, and dark,
Few hear our secret songs at night,
How do we live our lives without light?"

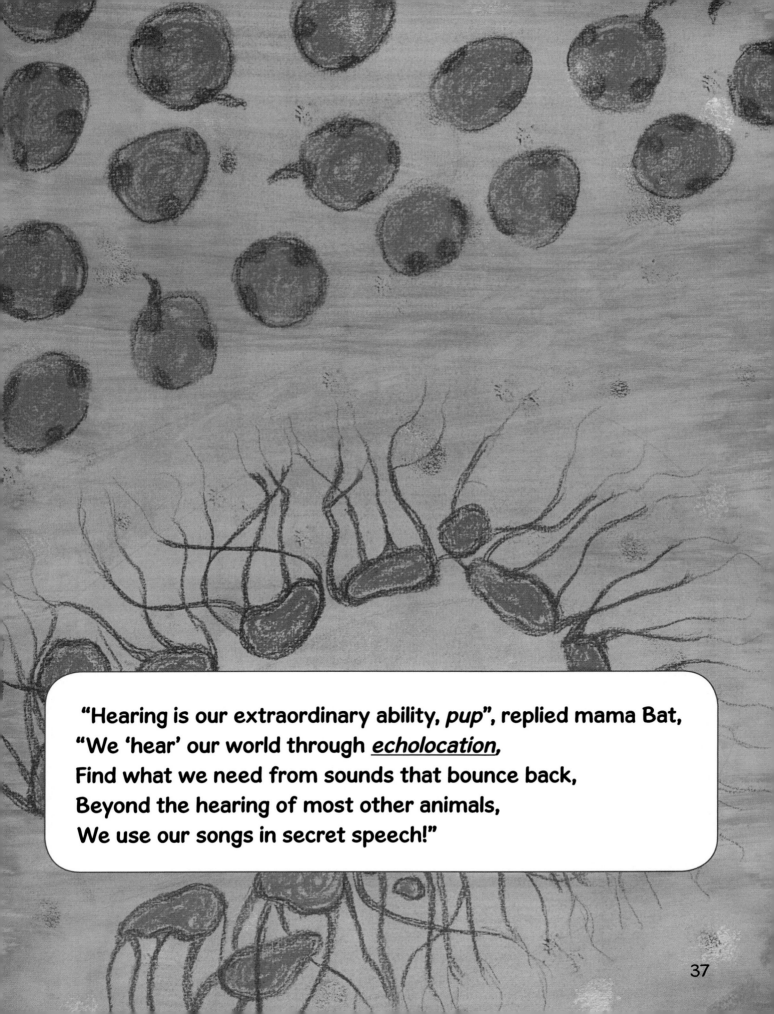

"Hearing is our extraordinary ability, *pup*", replied mama Bat,
"We 'hear' our world through <u>echolocation,</u>
Find what we need from sounds that bounce back,
Beyond the hearing of most other animals,
We use our songs in secret speech!"

EXPANDED GLOSSARY

1) **_Olfactory Bulb_** – Chemical compositions of the world around us enter our nose and the scent molecules activate receptors. These receptors send appropriate signals to the olfactory bulb, the first step in information processing in the brain. Olfaction, or chemical sensory perception, is also considered evolutionarily the "oldest" sense. Olfaction is often considered unique amongst the traditional 5 senses since information enters the brain directly and can evoke powerful emotions and behaviors. Information from the other 4 are extensively processed by the brainstem (near the spinal cord, evolutionarily the oldest part of the brain) and mid brain (connecting the brain stem to the rest of the brain) before entering the "neo cortex" and decision making part of the brain [4]. One of the first ever published images of nerve cells came from Golgi, based on the olfactory bulb [5]. Recently, interest in understanding olfaction has increased given the loss of the sense of smell – or anosmia – seen in many patients with COVID-19, and their subsequent loss of appetite [6]. Engineers are also interested in developing artificial nose-like sensors for more accurate sensing of toxic chemicals by understanding and emulating natural olfactory processes [7].

2) **_Bioluminescence_** – The chemical processes by which biological organisms generate light is called Bioluminescence. Although mostly seen in species that live in the twilight zone – or deep oceans where sunlight never penetrates – some organisms like fireflies also generate light to be used as communication signals. Bobtail squids enter a symbiotic (mutually beneficial) relationship with bioluminesent bacteria *Vibrio fischeri*. As a result, these squids are able to see in the dark, and provide

the bacteria a safe home within a light organ in an otherwise dangerous environment. These bacteria also help camouflage bobtail squid in a process called counterillumination. Interestingly, bobtail squid hatchlings and these bacteria find each other using unique chemical based signaling [8]. Other squids who live in the dark have evolved different visual adaptations – the giant squid has evolved to have eyes the size of footballs, uniquely suited to detect large predators like Sperm Whales [9]. Vision is one of the main senses using which we humans map the world around us and information from both eyes are processed to determine where objects are in space relative to us as well as what these objects are [10].

3) _**Sound Localization**_ – The ability to identify the source of a sound, in both direction and distance, is sound localization. Perceiving continously changing auditory stimuli presents a number of unique challenges to the brain. Auditory systems within various species, like the visual, have evolved to specifically solve unique challenges and behavioral needs faced by each. Accurate sound localization is one such problem. Sensory systems tasked with solving this computational problem need to compare incoming sounds from 2 sources – ususally the two eardrums – at the millisecond (1/1000th of a second) level. The sources of sounds that arrive a few milliseconds earlier, and slightly louder, in one ear than the other are closer to that side. Many species, including owls and humans who are able to localize sources of sounds, also have ears that are assymetric along the vertical plane that allows brains to detect when sounds originate from above or below their heads. In owls, the ear openings, hidden behind the preaural flaps in the ruff of feathers around the left and

right eyes (facial ruff) are assymmetric [11]. Many owl species combine a highly precise sound localization system with the ability to turn their heads about 270 degrees which helps them localize sounds all around them with high accuracy. Research has also shown that human auditory perceptual and memory is incredibly robust, with acoustic features as short as 10 ms being stored in long-term memory [12].

4) *Setae* – Setae are stiff hair-like bristles that in spiders comprise several categories of external sensory organs. Setae specialized as mechanosensory respond to mechanical stimuli such as touch, vibration, and air flow, chemosensory setae detect chemicals, and other setae sense enviromental fluctuations such as changes in temperature and humidity [13]. Many spiders are 'ambush predators' and use a combination of their web/silk and their mechanosensory system together to track and trap their prey. In essence, the webs they weave are extensions of their sensory system - physical displacement of the strands of silk are detected as tactile or touch signals. They rely extensively on this system to survive, and work efficiently to repair any damage to their webs. Some spiders can also vary the optical/ reflective properties of the web and its structure based on their habitat and the direction of light [14, 15]. While many spiders weave 'sticky' webs that prey get stuck to and have evolved specialized anti-adhesion traits so they don't get stuck to their own web [16], others weave non-sticky silk mats and use them as a sensory aid to track prey and use dry-adheshion techniques to prevent falling off the web [17]. Either way, by extending out a wide 'net', spiders are able to detect distance and direction of any movements on the web. Setae are also

used defensively sometimes – when threatened, tarantulas projectile-spray the setae on their legs at any attacker. There is a wide diversity in how different animals have evolved to detect physical stimuli of touch. In addition, the distribution of these mechanical somatosensors vary along the body based on evolutionary and behavioral needs. In humans, for instance, the fingertips and face have more touch receptors and are therefore more sensitive. Interestingly, by non-invasively researching the somatosensory "body maps" in infants, researchers are able to study how we start differentiating the "self" from "other" [18], leading to a strong sense of self as adults due to these somatotopic representations [19].

5) **_Taste Buds_** – Clusters of chemo-sensory receptors in the oral cavity of animals are called taste buds. They survey the chemical content of food for nutrients and toxicants. A wide variety of recptors detect these chemicals and convey taste perceptions such as sweet, bitter, umami, sour, and salty, while others have been shown to even detect carbohydrates and fats. Genetic variations between these receptors have often been linked to individual taste preferences such as some people not liking cilantro or brussels sprouts [20]. Another subset of receptors also sense chemicals that respond to hot spice (capsaicin) and mediate pain signals when dysfunctional or over-stimulated (see also 9, heat sensing in a different modality that uses the same mechanism at the molecular level) [21]. In addition, the thresholds of these receptors change with the temperatures of food, so if you ever wonder why food tastes different when hot/cold, now you know why! In humans, the overall experience of eating is therefore mediated by the texture (through mechanical

receptors), flavor (combination of all the individual tastants), smell (through olfactory sensation), temperature, and chemical irritants present in food [22]. Some species of fish have a highly developed sense of taste, with catfish and carps like goldfish dedicating about 20% of their brains to taste perceptions. One reason for this is that these fish have a large number of taste buds distributed outside their bodies in addition to those inside the oral cavity; the external taste buds let these fish orient and swim towards sources of food [23]. Inside the oral cavity, goldfish have specializations that allow them to sort food from stones, an ability which lets them search for food along the bottom of the sea/aquarium tank, called benthic feeding [24].

6) *__Carotid Bodies__* – Carotid artery is the blood vessel in the neck that carries blood from the heart to the brain. The carotid body is a sensory organ adjascent to the carotid artery. It contains chemosensors that detect chemical changes in the body by monitoring arterial blood for gas levels, especially Oxygen and Carbon Dioxide, pH, and temperature [25]. Carotid Bodies play a crucial role in the diving reflex – the coordinated response to being in water (in animals with lungs). In penguins, the carotid bodies detect the changes in blood gas composition during extended dive periods and signal the brain to change temperatures in some parts of the body to be lower using thermoregulation, redistribute oxygen between different internal organs, and switch to anaerobic metabolism (or maintain body functions in the absence of oxygen at the cost of accumulating toxins) as a last resort [26, 27]. Carotid bodies play a critical role in informing the nervous system, maintaining homeostasis in the respiratory and cardiovascular systems, as well as initiating immune and

metabolic responses to chemical changes in the blood. As a result, the carotid body has recently garnered interest as a potential target for therapy in disorders of these systems such as hypertension, hypoxia, and obesity [28-30].

7)   ***Neuromuscular Spindles*** – Neuromuscular spindles are types of stretch receptors present within skeletal muscles. Stretch receptors, also known as proprioceptors, are a diverse group of mechanosensory cells present in the body - within joints, muscles, tendons, and around internal organs. They sense the stretch/ movement/load/position of the joint/muscle/tendon/organ they are attached to, respectivley. In essence, proprioceptors monitor internal environments and mediate reflexive, or automatic, movements in nearly all motile animals [31]. In practical terms, these sensors inform brains of the speed and stretch of moving muscles (muscle spindles) as well as the load carried by joints and their limits (Golgi Tendon organs and Pacinian corpuscles), both of which require continuous monitoring and responses from the motor system. While sensing internal environments, these receptors indirectly affect how animals move and interact with objects in their environment, and as we get older, deterioration of proprioception affects posture, balance (see also 8, the role of the vestibular system in maintaining balance), movement, and muscle control. For this reason, scientists have focused on understanding the role of proprioceptors in aging [32]. Numerous balance and joint exercises have been shown to improve proprioception [33] to overcome the decline in their functions with injury or aging. In addition, mimicking proprioceptive sensing and reflexes have been a focus of robotics, with an emphasis

on building artificial systems capable of complex movements [34, 35].

8) **_Vestibular System_** – The vestibular system present in the inner ear senses the poisiton, movement, and orientation of the head. This information is used to compute a central estimate of head and body position, as well as our motion in space, by the brain. It is one of the oldest sensory system in vertebrates (animals with a backbone), the evolution of which allowed vertebrates to detect (and subsequently re-direct) their movements in any environment. The vestibular system senses both linear acceleration in any direction (including gravitational) as well as angular acceleration in 3 perpendicular planes to detect rotational movements [36]. Similar to proprioception (see also 7, the role of proprioceptors in movement), the continous monitoring of our head position is implicated in maintaining physical stability and in corrective movements of the eyes and head to compensate for any perturbations to our balance [37]. While this sensory process is largely unconscious, dysfunctions in the vestibular system result in debilitating symptoms and a profound decrease in quality of life due to symptoms like dizziness/vertigo, nausea, and migraines [38]. As a result, several vestibular rehabilitation programs have been developed that capitalize on the inherent plasticity (flexibility) of the balance system which requires it to compensate for perturbations very quickly [39]. Given the recent interest in space travel and how the human body might adapt to modified gravitational forces on other planetary bodies in space, scientists are even studying how the vestibular system mediates various physiological functions in different gravitiational situations [40]!

9) **_Infrared Radiation_** – Infrared radiation simply means that the light emitted has a wavelength greater than the red end of the visible spectrum, but shorter than microwaves. This form of light is emitted by heated objects. Any receiver of this radiation has to retain a stable temperature while sensing or risk overheating. Being able to percieve light that other organisms cannot lends a big advantage to biological, and more recently, bio-inspired man-made systems which can. Artificial infrared sensing systems span applications ranging from military (night vision, surveillance, target tracking), to medicine (non-invasive imaging including tumor imaging since tumors are more metabolically active and warmer than surrounding tissues) [41]. Pit organs in snakes have evolved to be highly sensitive to small differences in temperature, enabling them to map the body heat of living organisms and differentiate them from inorganic materials of similar size, at night. The anatomy of pit organs are unique as they provide sensitivity, detection distance of over 50 cm, and have inbuilt cooling mechanisms to prevent overheating [42]. While the heat sensing channels – the TRP family of receptors – are ubiquitous (widespread) in nature and evolutionarily conserved across various species (see also 5, heat-related pain and taste responses in humans) [21, 43], the pit organ type adaptation is unique in mapping and utilizing this information to inform hunting and other behaviors. They thus continue to inspire researchers and engineers to emulate their design in man-made infrared sensing applications.

10) **_Sensilla_** – Worms have simplified sensory receptors on the skin (epidermis), typically in the form of a hair- or rod- shape protrutions called sensilla. Worms are invertebrates (animals

without a backbone) of three types: flatworms (*Platyhelminthes* like tapeworms), roundworms (*Nematoda* like *C. elegans*), or segmented worms (*Annelida* like earthworms), so they span 3 taxonomic phyla (within the hierarchical system of categorizing living things based on similarity, phylum falls under kingdom). Worms have the highest concentration of sensory receptors in the head above the mouth, close to their ganglia (clusters of brain cells; primitive organisms have centralized ganglia that serve as a proto - or primitive - brain). They also have sensory cells near the tail region and the mid section [44]. Worms are found in extraordinarily diverse environments and despite a relatively simple nervous system that is conserved across various species, this system has adapted to sense a wide variety of environmental factors. Worm sensilla have been shown to detect olfactory/chemical, light, sound, and temperature cues [45-47]. Deep-sea worm species that live in extreme environments close to hydrothermic vents have evolved to have multiple layers protecting sensory cells so they can still detect changes in their unique and volatile environment [48]. Due to its small size and the ease of mapping its entire brain with just 302 neurons (specialized brain cells that transmit information) in the adult, 60 of which are sensory neurons, the worm *C. elegans,* has been extensively studied in neuroscience. Despite the simplicity of their nervous system, scientist continue to study worms and their ganglia since they shed light on how simple systems can perceive, learn, remember, and even how these process change with certain diseases [49].

11) **_Pheromones_** – Chemicals of biological origin that are used as precise and directed communication signals are called

pheromones. In social insects like honeybees, pheromones are critical for most functions including mating, foraging, tending the young, maintaining, and defending the hive. Honeybee communities have been studied extensively as 'superorganisms', since individuals work together altruistically for the 'good of the whole', and activities are coordinated without any decision centers [52]. Honeybees produce long-term (primer pheromones) and short-term (releaser pheromones) effectors, with multiple glands in the abdomen, feet, and head producing a variety of pheromones mediating development and behaviors in others. A newly hatched larva's development into worker, drones, or even queen depends on the pheromones they are exposed to at crititcal periods of development. Once pheromones are detected by the antennae of the recipient bee, they are relayed to the antennal lobe and subsequently the brain in the bee, similar to the olfactory bulb in mammals [50] (see also 1, olfactory bulb). Pheromones can even aggravate rebellion during regime changes in the hive – when the queen is replaced by a daughter – by alerting individuals to a decrease in genetic relatedness to others in the hive [51]. Beyond this type of communication, worker bees responsible for foraging for food and maintenance of the hive are able to asses continously evolving needs of the hive using pheromone, flight, and vibration (honeybee dance) signals [53, 54]. Superorganisms type of insect societies emerged about 100 million years ago and contain intact individuals within a functional whole, providing evolutionary biologists with a unique window into how multi-cellular organisms may have evolved from unicellular organisms about 700 million years ago [52]!

12) **_Salinity detector_** – Salinity detectors measure the conductivity, or amount of dissolved salts present in a solution. The mysteries of the narwhal tusk have fascinated scientists since the 1400s and its hypothesized functions have ranged from fighting to digging. Despite the lack of enamel (as seen in the white outer shell of human teeth), the narwhal tusk contains cementum, dentine, and pulpal tissue, consistent with being a tooth. A network of tubules in the dentin along the full length of the tooth were discovered, these tubules radiating outward from the core [55]. If you've eaten ice-cream too quickly, you'll know that teeth are sensory organs that detect high and low temperatures of food and can convey pain signals. Teeth have additionally been shown to detect pressure changes, chemical gradients and vibrations [56]. Neurophysiology, gene profiling, and heart rate experiments on the narwhal have confirmed that at least one role of their teeth lies in their sensory ability to distinguish high-salt and fresh-water solutions. The internal tubular structure of the tooth makes it resistant to fractures while maintaining flexibility. Since salinity of a solution dictates its pressure/density and temperature, the role of the narwhal tooth may be to detect rapid changes in water temperatures, essential to survival in the cold arctic waters where they live and need to come up for air [55]!

13) **_Polarized Light_** – While several animals are able to perceive visual information in color, butterflies and some species of insect pollinators have the broadest range and highest discriminability between colors in the animal kingdom! Furthermore, butterfly species vary in the type, number, and distribution of photoreceptors – light sensitive cells - within their

compound eyes. Compound eyes are composed of thousands of independent optical units (called Ommatidia) with lenses that detect light, so while the resolution - or the ability to distinguish individual features - is poor, animals with compound eyes have several advantages. These include a very wide field of view and high sensitivity to any movement across their visual field. Some species also sport intracellular lateral filtering pigments which add further nuance to the colors and patterns they are able to observe [57]. The tubular anatomy of individual optical units also results in compound eyes being able to detect polarized light, or the patterns in which light gets scattered by the atmosphere. To use this information for navigation, monarch butterflies and other insects have evolved to have enhanced sensitivity to the ultraviolet wavelength range within which polarization patterns in the sky have the greatest contrast [58]. Some butterflies have evolved to capitalize on having this unique perceptual ability by developing wing patterns that are only detectable in the ultraviolet range for species-specific communication [59]. Interestingly, some animlas with simple eyes - the type of eyes that humans have - have also evolved to detect ultraviolet light by using multi-layered retinas (see also the neuroscience image in 2, retina) to detect wavelengths beyond the visible spectrum - jumping spiders are able to assess the distance between themselves and different objects by assessing the amount of "ultraviolet blur" since objects that are closer appear fuzzier along the ultraviolet wavelength range [60].

14) *Ampullae of Lorenzini* – Marine sharks, skates, and rays contain pores on their skin surface with receptors highly sensitive to weak electric fields called Ampullae of Lorenzini. These can possibly

also detect magnetic fields which would help animals oreint to the earth's magnetic field. Few species have evolved the ability to detect electric fields, including monotremes (like platypus), sharks, rays, fresh water catfish, and electric eels; these animals capitalize on detecting weak electric fields produced by living organisms and use it to detect prey and to navigate in their environment. In addition to passive sensing, electric fish and eels employ active electrolocation by sending out pulses or waves of electricity to stun their prey and to actively map their environment based on electric fields [61]. While ampullae of lorenzini in fresh water and aquatic invertebrates evolved from lateral line receptors (see also 15, Neuromasts for more about lateral line receptors), electroreceptors in mammals like platypus seem to have evolved from the trigeminal somatosensory system (refer to the information about the neuroscience drawing in 4, as well as touch perception in humans), demonstrating convergent evolution (where diverse species have converged to similar abilities or behaviors) [62, 63]. A recent bioengineering endeavor has been to use electric fields to construct and characterize artificial tissues from their component cells using electric fields [64].

15) **_Neuromasts_** – Neuromasts are sensory organs present on the skin surface or within canals (invaginations) in the skin, along the head and body of amphibians and fishes [62]. Neuromasts are a type of lateral line receptors (see also 14, ampullae of lorenzini, which also evolved from lateral line receptors) since they line up along the body plan of the organism, which are believed to be the evolutionary predecessor for both [65]. They are mainly sensitive to water currents, but also show sensitivity to low

frequency vibrations which lets them detect certain sounds under water since sounds move as pressure waves through liquids [66]. Since many species of frogs also use sounds for within-species communication, some frogs have also evolved a tympanic membrane (ear-drum) behind their eyes, while others can hear with the help of their lungs [67]. Essentially, neuromasts are mechanical sensors that are present along the surface of animals and help them generate a map of their surroundings based on water currents and vibrations, similar to how humans create visual maps of our world [68]. Recent research has also demonstrated that activating the lateral line receptors in tadpoles evoked fast motor responses (eg., turning away from dangerous currents), suggesting that these receptors are important for survival [69]. Interestingly, cave dwelling fish have evolved to have numerous neuromasts to successfully find food in the dark by evolving 'Vibration Attraction Behavior' not seen in other fish [70]. Understanding the patterns of neuromast development and plasticity is an active area of research which often sheds light on the processes by which behavioral needs shape adaptations, and eventually, evolution.

16) *Echolocation* – Using reflected sounds to map the location of objects is echolocation. In nature, very few species like dogs, dolphins, and bats have evolved to hear sounds beyond the hearing range of most other species. All of them use this sensory ability to both communicate and move through the world. In order to guage reflected sounds, bats generate airborne species-specific ultrasonic calls which bounce off objects in their environment and return to them. They are even able to modify their call structure mid-flight based on flight speed and

distance to objects in their auditory world [71]. In order to hear and process high frequency – or ultra- sounds, the auditory system in bats has evolved several modifications to the cochlea (hearing organ; see also neuroscience drawing in 3) as well as other areas of the brain processing auditory information [72, 73]. They also have brain cells dedicated to navigating and keeping track of their angle and distance from targets during flight [74]. Bats communicate with each other using high freqeuncy chirps, screeches, and songs, and this has fascinated scientists since the discovery that bat infants 'babble' similar to human infants. Babbling is considered a milestone stage in language development in humans, suggesting that other mammals with vocal communication might have similar developmental mechanisms [75]. Interestingly, the emergence of submarine warfare during World War I led to humans developing 'SONAR' (sound navigation and ranging) technology to locate underwater objects using sounds [76]. Today, ultrasound technologies are widely used in science, research, medicine, geo-spatial mapping, and robotics, to name a few.

ACKNOWLEDGEMENTS

I am deeply indebted to the incredible mentors who continue to inspire me in my neuroscientific pursuits, especially Dr. Simon Thorpe & Dr. Florence Rémy-El Boustani (Centre National de la Recherche Scientifique, or CNRS, France), as well as Dr. Suzana Petanceska & Dr. Lorenzo Refolo (National Institute on Aging, USA). It was my privilege to complete a doctorate with Simon and Florence as mentors. Simon, I have left every meeting with you feeling stimulated, enthusiastic, uplifted, and full of ideas. Thank you for always encouraging my desire to look at the bigger picture. Florence, thank you for teaching me the principles of doing "good science" – I carry the lessons I learned from you every day. Suzana and Larry, I am very fortunate that I get to work with both of you; your tireless endeavors to make scientific pursuits more rigorous, robust, open, reproducible, and the research environment more supportive and conducive to exemplary science inspire me every day.

Thank you also to past mentors Dr. Suvarna Alladi (National Institute on Mental Health and Neurosciences, or NIMHANS, India) and Dr. Jason Barton (University of British Columbia, Canada) for introducing me to the world of Neuroscience. Thank you to Dr. Srikanth Raghavendran (Sastra University, India) for showing me how much fun and integral Mathematics is in Biology. Thank you to Dr. Shihab Shamma & Dr. Pingbo Yin (University of Maryland, College Park, USA) for introducing me to the wonderful world of Neurophysiology and quantifying animal behavior.

This book would not have been possible without the incredible Tuesday Night Drawing Group in Georgetown, Washington, D.C. For over 30 years, artists in the DMV (DC-Maryland-Virginia) area have been meeting up once a week to draw, paint, sculpt, and discuss art. Throughout the pandemic, this group inspired and encouraged me to keep working on my various science-art projects. I am particularly grateful to Micheline Klagsbrun, who creates transformative works inspired by nature, Dr. Marc Allen and Dr. Lucy McFadden, fellow scientists (astrophysicists) and artists, and Irene Pantelis, an artist who highlights environmental issues. I'm also grateful to Dr. Aneta Shine, Nancy Frank, Renato Salazar, Barry Goodman, and Sam Freeman. Thank you to my French art teacher, Sandrine Follère, who introduced me to a whole new world when she taught me to channel intense emotions into art.

During my engineering and neuroscientific career, I've lived in 4 and visited over 20 countries. This would not have been possible, or fun, without my incredible friends and colleagues from over the years. Although too many to name everyone individually, I am indebted to you all, and you inspire me every day. Thank you to my friends in Vancouver, in Toulouse, and in Paris. Thank you to all my friends from my undergraduate, graduate student, and post-doctoral years who now live all over the world. You've been my family-away-from-home and have supported me through thick and thin. These were periods where I grew into myself, and the adventures we've had have been unforgettable. Thank you to all my wonderful friends and colleagues in the USA and particularly the DMV area – I fell in love with and stay here because of friends like you.

Thank you to the volunteer coordinators at Smithsonian National Museum of Natural History for selecting me as an insect zoo and butterfly pavilion ambassador. Spending time in your environment of science, learning, and around amazing arthropods has been a delight.

Thank you also to Reading Partners for the opportunity to volunteer with you reading to kids. Volunteering with kids through the pandemic was a source of joy during a difficult time.

Last, but definitely not least, I would like to thank my wonderful parents Dr. Viswanathan Poyyamani Swaminathan & Mrs. Mahalakshmi Viswanathan from the bottom of my heart. Thank you for always encouraging me to follow my dreams, even when it made life downright difficult for you to do so. You've instilled in me a spirit of public service, a zest for social justice, and passion for learning new things. I always turn to your wisdom and philosophy to guide me through life. Despite having the gift of gab, I am at a loss to describe the depth of my gratitude; for always showing me by example how to do the right thing even when it is difficult, for inspiring me to be a better person every day, and for teaching me to be curious about the world.

REFERENCES

References in Text

1. M, C.G., *The Cell: A Molecular Approach*. 2nd edition ed. 2000, Sunderland (MA): Sinauer Associates.

2. Carpenter, S.K., *Some Neglected Contributions of Wilhelm Wundt to the Psychology of Memory*. Psychological Reports, 2005. **97**(1): p. 63-73.

3. DeFelipe, J., *The Evolution of the Brain, the Human Nature of Cortical Circuits, and Intellectual Creativity*. Frontiers in Neuroanatomy, 2011. **5**.

4. Purves D, A.G., Fitzpatrick D, et al.,, *The Organization of the Olfactory System.*, in *Neuroscience*. 2001: Sunderland (MA).

5. Shepherd, G.M., et al., *The first images of nerve cells: Golgi on the olfactory bulb 1875*. Brain Research Reviews, 2011. **66**(1-2): p. 92-105.

6. de Melo, G.D., et al., *COVID-19 - related anosmia is associated with viral persistence and inflammation in human olfactory epithelium and brain infection in hamsters*. Science Translational Medicine, 2021. **13**(596): p. eabf8396.

7. Sankaran, S., L.R. Khot, and S. Panigrahi, *Biology and applications of olfactory sensing system: A review*. Sensors and Actuators B: Chemical, 2012. **171-172**: p. 1-17.

8. Zink, K.E., et al., *A Small Molecule Coordinates Symbiotic Behaviors in a Host Organ*. mBio, 2021. **12**(2): p. e03637-20.

9. Nilsson, D.-E., et al., *A Unique Advantage for Giant Eyes in Giant Squid*. Current Biology, 2012. **22**(8): p. 683-688.

10. Goodale, M.A., et al., *A neurological dissociation between perceiving objects and grasping them*. Nature, 1991. **349**(6305): p. 154-156.

11. Knudsen, E.I. and M. Konishi, *Mechanisms of sound localization in the barn owl (Tyto alba)*. Journal of Comparative Physiology ? A, 1979. **133**(1): p. 13-21.

12. Viswanathan, J., et al., *Long Term Memory for Noise: Evidence of Robust Encoding of Very Short Temporal Acoustic Patterns*. Frontiers in Neuroscience, 2016. **10**.

13. Schacht, M.I., M. Francesconi, and A. Stollewerk, *Distribution and development of the external sense organ pattern on the appendages of*

postembryonic and adult stages of the spider Parasteatoda tepidariorum. Development Genes and Evolution, 2020. **230**(2): p. 121-136.

14. Zschokke, S., S. Countryman, and P.E. Cushing, *Spiders in space—orb-web-related behaviour in zero gravity.* The Science of Nature, 2021. **108**(1).

15. Kane, D., et al., *Orb web spider silks: how their optics affects potential visibility.* ANZCOP. Vol. 11200. 2019: SPIE.

16. Briceño, R.D. and W.G. Eberhard, *Spiders avoid sticking to their webs: clever leg movements, branched drip-tip setae, and anti-adhesive surfaces.* Naturwissenschaften, 2012. **99**(4): p. 337-41.

17. Poerschke, B., S.N. Gorb, and C.F. Schaber, *Adhesion of Individual Attachment Setae of the Spider Cupiennius salei to Substrates With Different Roughness and Surface Energy.* Frontiers in Mechanical Engineering, 2021. **7**.

18. Marshall, P.J. and A.N. Meltzoff, *Body maps in the infant brain.* Trends in Cognitive Sciences, 2015. **19**(9): p. 499-505.

19. Craig, A.D., *The sentient self.* Brain Structure and Function, 2010. **214**(5-6): p. 563-577.

20. Roper, S.D. and N. Chaudhari, *Taste buds: cells, signals and synapses.* Nat Rev Neurosci, 2017. **18**(8): p. 485-497.

21. Aroke, E.N., et al., *Taste the Pain: The Role of TRP Channels in Pain and Taste Perception.* International Journal of Molecular Sciences, 2020. **21**(16): p. 5929.

22. Green, B.G., *Heat as a Factor in the Perception of Taste, Smell, and Oral Sensation.*, in *Nutritional Needs in Hot Environments: Applications for Military Personnel in Field Operations.*, I.o.M.U.C.o.M.N.R.M. BM, Editor. 1993, National Academies Press (US): Washington (DC).

23. Caprio, J., et al., *The taste system of the channel catfish: from biophysics to behavior.* Trends in Neurosciences, 1993. **16**(5): p. 192-197.

24. Finger, T.E., *Sorting food from stones: the vagal taste system in Goldfish, Carassius auratus.* J Comp Physiol A Neuroethol Sens Neural Behav Physiol, 2008. **194**(2): p. 135-43.

25. Forbes J, M.R., *Anatomy, Head and Neck, Carotid Bodies.* Updated 2021 Aug 1, StatPearls Publishing: Treasure Island (FL).

26. Butler, P.J., *The Diving Reflex in Free-Diving Birds*, in *Arctic Underwater Operations: Medical and Operational Aspects of Diving Activities in Arctic Conditions*, L. Rey, Editor. 1985, Springer Netherlands: Dordrecht. p. 49-61.

27. P.J Ponganis, G.L.K., *Diving physiology of birds: a history of studies on polar species.* Comparative Biochemistry and Physiology Part A: Molecular & Integrative Physiology, 2000. **126**(2): p. 143-151.

28. Badoer, E., *The Carotid Body a Common Denominator for Cardiovascular and Metabolic Dysfunction?* Front Physiol, 2020. **11**: p. 1069.

29. Conde, S.V., J.F. Sacramento, and F.O. Martins, *Immunity and the carotid body: implications for metabolic diseases.* Bioelectronic Medicine, 2020. **6**(1).

30. Leonard, E.M., S. Salman, and C.A. Nurse, *Sensory Processing and Integration at the Carotid Body Tripartite Synapse: Neurotransmitter Functions and Effects of Chronic Hypoxia.* Front Physiol, 2018. **9**: p. 225.

31. Tuthill, J.C. and E. Azim, *Proprioception.* Current Biology, 2018. **28**(5): p. R194-R203.

32. Ferlinc, A., et al., *The Importance and Role of Proprioception in the Elderly: a Short Review.* Mater Sociomed, 2019. **31**(3): p. 219-221.

33. Aman, J.E., et al., *The effectiveness of proprioceptive training for improving motor function: a systematic review.* Front Hum Neurosci, 2014. **8**: p. 1075.

34. Hannaford, B., K. Jaax, and G. Klute, *Bio-Inspired Actuation and Sensing.* Autonomous Robots, 2001. **11**(3): p. 267-272.

35. Liu, X., et al., *Robotic investigation on effect of stretch reflex and crossed inhibitory response on bipedal hopping.* Journal of The Royal Society Interface, 2018. **15**(140): p. 20180024.

36. Highstein, S.M., *Anatomy and Physiology of the Central and Peripheral Vestibular System: Overview.* 2004, Springer-Verlag. p. 1-10.

37. Jones, S.M., et al., *Anatomical and Physiological Considerations in Vestibular Dysfunction and Compensation.* Semin Hear, 2009. **30**(4): p. 231-241.

38. Agrawal, Y., B.K. Ward, and L.B. Minor, *Vestibular dysfunction: prevalence, impact and need for targeted treatment.* J Vestib Res, 2013. **23**(3): p. 113-7.

39. Han, B.I., H.S. Song, and J.S. Kim, *Vestibular rehabilitation therapy: review of indications, mechanisms, and key exercises.* J Clin Neurol, 2011. **7**(4): p. 184-96.

40. Morita, H., et al., *Understanding vestibular-related physiological functions could provide clues on adapting to a new gravitational environment.* The Journal of Physiological Sciences, 2020. **70**(1).

41. Shen, Q., et al., *Bioinspired infrared sensing materials and systems.* Advanced Materials, 2018. **30**(28): p. 1707632.

42. Moon, C., *Infrared-sensitive pit organ and trigeminal ganglion in the crotaline snakes.* Anatomy & Cell Biology, 2011. **44**(1): p. 8.

43. Caterina, M.J., et al., *The capsaicin receptor: a heat-activated ion channel in the pain pathway.* Nature, 1997. **389**(6653): p. 816-824.

44. Kiszler, G., et al., *Organization of the sensory system of the earthworm<i>Lumbricus terrestris</i>(Annelida, Clitellata) visualized by DiI.* Journal of Morphology, 2012. **273**(7): p. 737-745.

45. Bumbarger, D.J., et al., *Three-dimensional reconstruction of the amphid sensilla in the microbial feeding nematode, Acrobeles complexus (nematoda: Rhabditida).* The Journal of Comparative Neurology, 2009. **512**(2): p. 271-281.

46. Ward, A., et al., *Light-sensitive neurons and channels mediate phototaxis in C. elegans.* Nat Neurosci, 2008. **11**(8): p. 916-22.

47. Iliff, A.J., et al., *The nematode C. elegans senses airborne sound.* Neuron, 2021. **109**(22): p. 3633-3646.e7.

48. Shigeno, S., et al., *Sensing deep extreme environments: the receptor cell types, brain centers, and multi-layer neural packaging of hydrothermal vent endemic worms.* Frontiers in Zoology, 2014. **11**(1).

49. *C. elegans II.* 2 ed. C. elegans II, ed. D.L. Riddle, et al. 1997, Cold Spring Harbor (NY): Cold Spring Harbor Laboratory Press Copyright © 1997, Cold Spring Harbor Laboratory Press.

50. Bortolotti, L., Costa, C., *Chemical Communication in the Honey Bee Society.*, in *Neurobiology of Chemical Communication.*, C. Mucignat-Caretta, Editor. 2014, CRC Press/Taylor & Francis: Boca Raton (FL).

51. Woyciechowski, M. and K. Kuszewska, *Swarming Generates Rebel Workers in Honeybees.* Current Biology, 2012. **22**(8): p. 707-711.

52. Seeley, T.D., *The Honey Bee Colony as a Superorganism.* American Scientist, 1989. **77**(6): p. 546-553.

53. Boucher, M. and S.S. Schneider, *Communication signals used in worker-drone interactions in the honeybee, Apis mellifera.* Animal Behaviour, 2009. **78**(2): p. 247-254.

54. Camazine, S., *The regulation of pollen foraging by honey bees: how foragers assess the colony's need for pollen.* Behavioral Ecology and Sociobiology, 1993. **32**(4).

55. Nweeia, M.T., et al., *Sensory ability in the narwhal tooth organ system.* The Anatomical Record, 2014. **297**(4): p. 599-617.

56. Närhi, M., *The neurophysiology of the teeth.* Dent Clin North Am, 1990. **34**(3): p. 439-48.

57. Blackiston, D., A.D. Briscoe, and M.R. Weiss, *Color vision and learning in the monarch butterfly, Danaus plexippus (Nymphalidae).* Journal of Experimental Biology, 2011. **214**(3): p. 509-520.

58. Roger, *Polarization Vision: Drosophila Enters the Arena.* Current Biology, 2012. **22**(1): p. R12-R14.

59. Briscoe, A.D., et al., *Positive selection of a duplicated UV-sensitive visual pigment coincides with wing pigment evolution in Heliconius butterflies.* Proceedings of the National Academy of Sciences, 2010. **107**(8): p. 3628-3633.

60. Institution, S., *Micro Life: Miracles of the Miniature World Revealed.* 2021: Dorling Kindersley Publishing.

61. Von Der Emde, G., *Electric Fields and Electroreception: How Electrosensory Fish Perceive Their Environment.* 2001, Springer Berlin Heidelberg. p. 313-329.

62. Bullock, T.H., et al. *Electroreceptors and Other Specialized Receptors in Lower Vertebrates.* in *Handbook of Sensory Physiology.* 1974.

63. Scheich, H., et al., *Electroreception and electrolocation in platypus.* Nature, 1986. **319**(6052): p. 401-402.

64. Markx, G.H., *The use of electric fields in tissue engineering: A review.* Organogenesis, 2008. **4**(1): p. 11-7.

65. Modrell, M.S., et al., *Electrosensory ampullary organs are derived from lateral line placodes in bony fishes.* Nature Communications, 2011. **2**(1): p. 496.

66. Higgs, D.M. and C.A. Radford, *The contribution of the lateral line to 'hearing' in fish.* Journal of Experimental Biology, 2012. **216**(8): p. 1484-1490.

67. Simmons, A.M., *Auditory neuroethology: What the frog's lungs tell the frog's ear.* Current Biology, 2021. **31**(7): p. R350-R351.

68. Pujol-Martí, J. and H. López-Schier, *Developmental and architectural principles of the lateral-line neural map.* Front Neural Circuits, 2013. **7**: p. 47.

69. Saccomanno, V., et al., *The early development and physiology of Xenopus laevis tadpole lateral line system.* Journal of Neurophysiology, 2021. **126**(5): p. 1814-1830.

70. Yoshizawa, M., et al., *Evolution of a behavioral shift mediated by superficial neuromasts helps cavefish find food in darkness.* Curr Biol, 2010. **20**(18): p. 1631-6.

71. Jones, G. and M.W. Holderied, *Bat echolocation calls: adaptation and convergent evolution.* Proceedings of the Royal Society B: Biological Sciences, 2007. **274**(1612): p. 905-912.

72. Covey, E., *Neurobiological specializations in echolocating bats.* The Anatomical Record Part A: Discoveries in Molecular, Cellular, and Evolutionary Biology, 2005. **287A**(1): p. 1103-1116.

73. Sulser, R.B., et al., *Evolution of inner ear neuroanatomy of bats and implications for echolocation.* Nature, 2022. **602**(7897): p. 449-454.

74. Abbott, A., *Sat-nav neurons tell bats where to go.* Nature, 2017.

75. Fernandez, A.A., et al., *Babbling in a vocal learning bat resembles human infant babbling.* Science, 2021. **373**(6557): p. 923-926.

76. D'amico, A. and R. Pittenger, *A brief history of active sonar.* 2009, SPACE AND NAVAL WARFARE SYSTEMS CENTER SAN DIEGO CA.

References for Neuroscientific Paintings

77. DeFelipe, J.R.n.y.C.S., *Cajal's butterflies of the soul : science and art.* 2010, Oxford; New York: Oxford University Press.

78. C., G., *Sulla fina struttura dei bulbi olfactorii. (On the fine structure of the olfactory bulb).* Rivista Sperimentale di Freniatria e Medicina Legale., 1875(1): p. 405–425.

79. Kopsch, F., *Mitteilungen ilber das Ganglion opticum der Cephalopode.* Int. Monats. Anat. Physiol., 1899. **16**: p. 3-24.

80. Swanson, L.W., L. King, and E. Himmel, *The beautiful brain: The drawings of Santiago Ramón Y Cajal.* The beautiful brain: The drawings of Santiago Ramón Y Cajal., ed. E.A. Newman, A. Araque, and J.M. Dubinsky. 2017, New York, NY, US: Abrams Books. 207-207.

81. deCastro, F., *Nota sobre ciertas terminaciones nerviosas en el ganglio cervical superior simpático humano. .* Trab. Lab. Invest. Biol., 1916. **19**: p. 241–340.

82. Morais, S., *The Physiology of Taste in Fish: Potential Implications for Feeding Stimulation and Gut Chemical Sensing.* Reviews in Fisheries Science & Aquaculture, 2017. **25**(2): p. 133-149.

83. deCastro, F., *Sur la structure et l'innervation de la glande intercarotidienne (glomus caroticum) de l'homme et des mammiféres, et sur un nouveau système d'innervation autonome du nerf glossopharyngien.* Trab. Lab. Invest. Biol. Univ. Madrid, 1926. **24**: p. 365–432.

84. Tello, J.F., *Genesis de las terminaciones nerviosas motrices y sensitivas. .* Trab. Lab. Invest. Biol. Univ. Madrid, 1917. **15**: p. 101-199.

85. Fitzakerley, J. *Semicircular canals.* 2015 2/7/2015 7/2/2022].

86. Bullock, T.H. and W.S. Fox, *The Anatomy of the Infra-red Sense Organ in the Facial Pit of Pit Vipers.* Journal of Cell Science, 1957: p. 219-234.

87. Cajal, S.R.Y., *Neuroglia y neurofibrillas del Lumbricus.* Trab. Lab. Invest. Biol. Univ. Madrid, 1904. **3**: p. 277-285.

88. Yan, X., et al., *Glomerular Organization of the Antennal Lobes of the Diamondback Moth, Plutella xylostella L.* Frontiers in Neuroanatomy, 2019. **13**.

89. Brown, S.M., et al., *Stereological Analysis Reveals Striking Differences in the Structural Plasticity of Two Readily Identifiable Glomeruli in the Antennal Lobes of the Adult Worker Honeybee.* The Journal of Neuroscience, 2002. **22**(19): p. 8514-8522.

90. Narendra, A., et al., *Compound Eye Adaptations for Diurnal and Nocturnal Lifestyle in the Intertidal Ant, Polyrhachis sokolova.* PLoS ONE, 2013. **8**(10): p. e76015.

91. Russell, I.J., *Amphibian Lateral Line Receptors.* 1976, Springer Berlin Heidelberg. p. 513-550.

92. Quinzio, S. and M. Fabrezi, *The Lateral Line System in Anuran Tadpoles: Neuromast Morphology, Arrangement, and Innervation.* The Anatomical Record, 2014. **297**(8): p. 1508-1522.

ABOUT THE AUTHOR

Dr. Jaya Viswanathan is a neuroscientist, engineer, and artist born and raised in India. She is passionate about understanding how the brain works, so after graduating with an engineering degree (with honors), she pursued a Masters in Neuroscience followed by a Doctorate in Cognitive Neuroscience. She then worked as a post-doctoral fellow for 4 years before joining the Division of Neuroscience at the National Institute on Aging as a Program Analyst [C]. Throughout her scientific career, she has worked on understanding how brains interpret the world around us. She is passionate about providing STEAM (Science, Technology, Engineering, Art, and Math) educational resources to all students regardless of social or economic inequalities. She has been featured in the "Lemme ask Ya" podcast hosted by Clay and Jake (Episode 17: Neuro-Knowledge, Not for Naught) available on Spotify, Apple Podcasts, YouTube, and Patreon. She fell in love with art when pursuing her Doctorate in France and paints every week. Dr. Viswanathan's Neuroscience art has been featured in the Air Force Research Laboratory's Basic Research Art of Science Showcase, 2022. She also regularly exhibits her art at local galleries in the DC/Maryland/Virginia area where she resides. Through her art and science, she aims to convey the underlying message of social awareness, appreciation, and acceptance of diversity in nature. Feel free to drop her a line at babysensesbook@gmail.com and stay tuned to updates about the book at https://babysensesbook.com/!

Further information and requests for full sized prints

This project was born in love. Love for the brain which I haven't been able to shake since the age of 14. Love for the incredibly complex world we live in with a rich diversity to be treasured. A fascination for understanding and love of mysteries of evolution – particularly convergent and divergent evolution. Love for my parents who have sacrificed so much for me. Last but not least, the wonders of behavior and love itself. For more information about the project, check out https://babysensesbook.com/.

Animal Paintings (in order)

1. Lemur (Order *Primates*) adult with cub. Endemic to Madagascar; Endangered conservation status (CS).
2. Bobtail Squid (Order *Sepiolida*) adult with *Vibrio fischeri,* and hatchling/ paralarva without bacteria. Found in shallow coastal waters off the Pacific, Atlantic, and Indian Oceans; Data Deficient CS.
3. Tawny Owl (*Strix aluco*) adult with chick. Habitat ranges from the UK to West Siberia; Least Concern CS.
4. Golden orb-weaver (Genus *Nephila*) adult female with spiderling. Found in warm regions around the world; Least Concern CS.
5. Butterfly telescope Goldfish (*Carassius auratus*) adult with fingerling. Originally from China, fancy goldfish species have been selectively bred for appearance traits over centuries; Least Concern CS.
6. Emperor Penguin (*Aptenodytes forsteri*) male adult with chick. Endemic to Antarctica; Near Threatened CS.
7. Silvery Langurs/Lutungs (Family *Cercopithecidae*) adult with child. Native to Southeast Asia; Vulnerable CS.

8. Pygmy Three-toed Sloth (*Bradypus pygmaeus*) adult with infant. Endemic to red mangroves on Isla Escudo de Veraguas; Critically Endangered CS.

9. Green Pit Viper (Genus *Trimeresurus*) female adult in nest. Native to Southeast Asia; Least Concern CS.

10. Earthworm (*Lumbricus terrestris*) adult with wormlet. Native to Western Europe, found widely distributed in temperate to mild boreal climates; Invasive species, Least Concernt CS.

11. Honeybee (Genus *Apis*) Queen in honeycomb with worker bee and larvae. Indigenous to Eurasia, domesticated species have been introduced in other places; Extinct/Data Deficient CS (in the wild).

12. Narwhal (*Monodon monoceros*) adult with calf. Lives year-round in Arctic waters around Greenland, Canada, and Russia; recently reclassified as Least Concern CS (from Near Threatened).

13. Monarch Butterfly (*Danaus plexippus*) adult with caterpillar. Commonly from North America, ranges worldwide; Least Concern CS for the species overall, but migratory subspecies recently classified as Endangered CS.

14. Bluespotted Ribbontail Ray (*Taeniura lymma*) adult stingray with pup. Common throughout the tropical Indian and western Pacific Oceans in coral-reef habitats; Least Concern CS.

15. Dyeing Poison Dart Frog (*Dendrobates tinctorius*) adult male with tadpoles. Endemic to eastern portion of South America (Venezuela, French Guiana, Guyana, Suriname, Brazil); Least Concern CS.

16. Bat (Order *Chiroptera*) adult with infant. Species found worldwide except very cold regions; CS is species specific.

Neuroscience Paintings (in order)

1. Olfactory bulb of the dog
2. Retina of the squid

3. Human Organ of Corti
4. Gasser ganglion
5. Tastebud of the fish
6. Carotid Body of the cat
7. Human Neuromuscular Spindle
8. Vestibular hair cell
9. Pit Organ of the snake
10. Ventral chain ganglion of earthworm
11. Antennal lobe of bee
12. Narwhal tooth
13. Compound Eye
14. Ampullae of Lorenzini
15. Lateral line receptor
16. Cochlear Nucleus of Bat

Print Requests

Go to https://babysensesbook.com/ for more information, to order digital full sized images of any of the paintings in this book as well as to learn more.

Printed in the United States
by Baker & Taylor Publisher Services